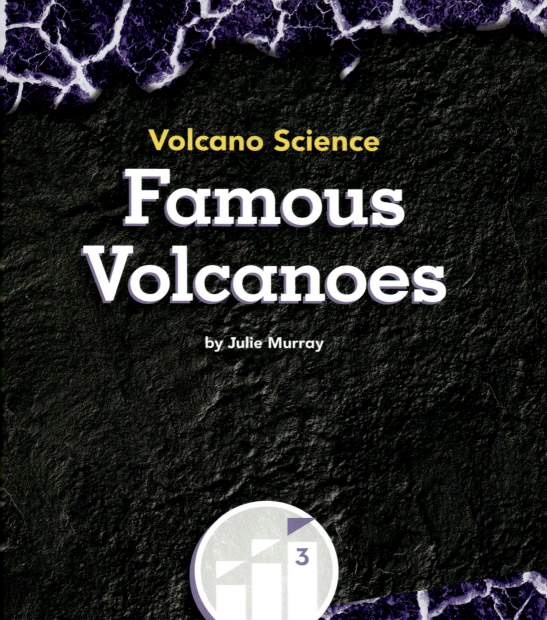

Volcano Science

# Famous Volcanoes

by Julie Murray

Dash!
LEVELED READERS
An Imprint of Abdo Zoom • abdobooks.com

### Level 1 – Beginning
Short and simple sentences with familiar words or patterns for children who are beginning to understand how letters and sounds go together.

### Level 2 – Emerging
Longer words and sentences with more complex language patterns for readers who are practicing common words and letter sounds.

### Level 3 – Transitional
More developed language and vocabulary for readers who are becoming more independent.

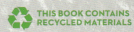

## abdobooks.com

Published by Abdo Zoom, a division of ABDO, PO Box 398166, Minneapolis, Minnesota 55439. Copyright © 2023 by Abdo Consulting Group, Inc. International copyrights reserved in all countries. No part of this book may be reproduced in any form without written permission from the publisher. Dash!™ is a trademark and logo of Abdo Zoom.

Printed in the United States of America, North Mankato, Minnesota.
052022
092022

Photo Credits: Getty Images, Shutterstock
Production Contributors: Kenny Abdo, Jennie Forsberg, Grace Hansen, John Hansen
Design Contributors: Candice Keimig, Neil Klinepier

### Library of Congress Control Number: 2021950305

### Publisher's Cataloging in Publication Data
Names: Murray, Julie, author.
Title: Famous volcanoes / by Julie Murray.
Description: Minneapolis, Minnesota : Abdo Zoom, 2023 | Series: Volcano science | Includes online resources and index.
Identifiers: ISBN 9781098228392 (lib. bdg.) | ISBN 9781098229238 (ebook) | ISBN 9781098229658 (Read-to-Me ebook)
Subjects: LCSH: Volcanoes--Juvenile literature. | Volcanic eruptions--Juvenile literature. | Volcanism--Juvenile literature. | Physical geography--Juvenile literature.
Classification: DDC 551.21--dc23

# Table of Contents

Famous Volcanoes . . . . . . . . . . . . . 4

Volcanoes Around
the World . . . . . . . . . . . . . . . . . . . 8

Volcanoes in the
United States . . . . . . . . . . . . . . . 16

More Famous Volcanoes . . . . . . . 22

Glossary . . . . . . . . . . . . . . . . . . . 23

Index . . . . . . . . . . . . . . . . . . . . . 24

Online Resources . . . . . . . . . . . . 24

# Famous Volcanoes

Earth's volcanoes have been erupting for billions of years. Today, there are more than 1,500 **active** volcanoes around the world.

Volcanoes can be **famous** for different reasons. Some are known for their explosive eruptions.

Others are known, sadly, for the amount of lives lost due to an eruption.

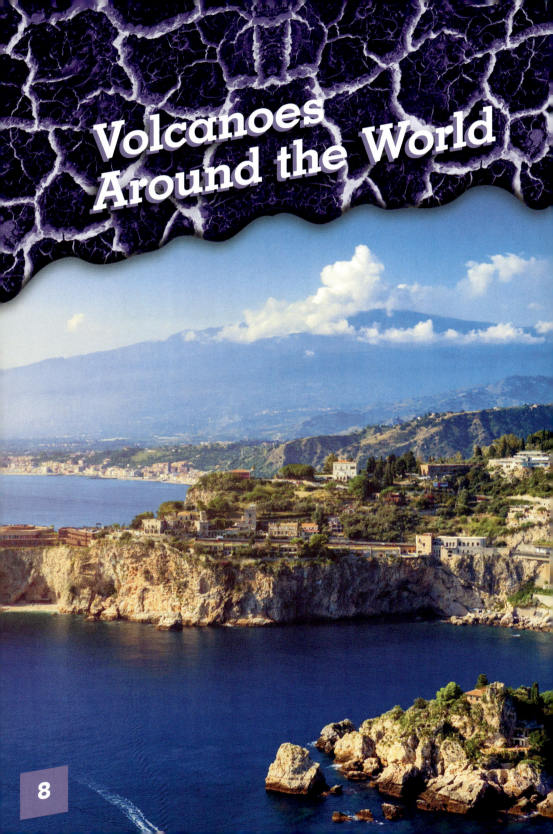
# Volcanoes Around the World

Mount Etna is one of the most **active** volcanoes in the world. It is in Sicily, Italy. It has been erupting for thousands of years.

Mount Pinatubo is in the Philippines. Its eruption in 1991 sent an **ash cloud** 22 miles (35 km) into the air. Hundreds of people died. It also left more than 200,000 people homeless.

Mount Tambora is in Indonesia. Its 1815 eruption was the most destructive volcanic event in **recorded history**. More than 120,000 people lost their lives.

Mount Vesuvius erupted in 79 CE. It buried the city of Pompeii and its people under 20 feet (6.1 m) of ash and **debris**. It remained that way for thousands of years. Today, much of the city has been uncovered.

# Volcanoes in the United States

Kīlauea is an **active shield volcano** on the Big Island in Hawaii. It has been continuously erupting since 1983. It is one of the longest eruptions ever recorded.

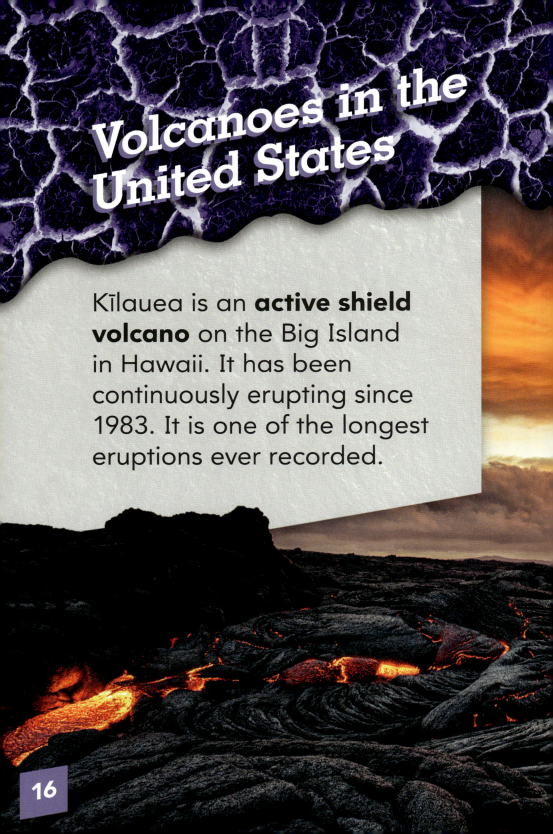

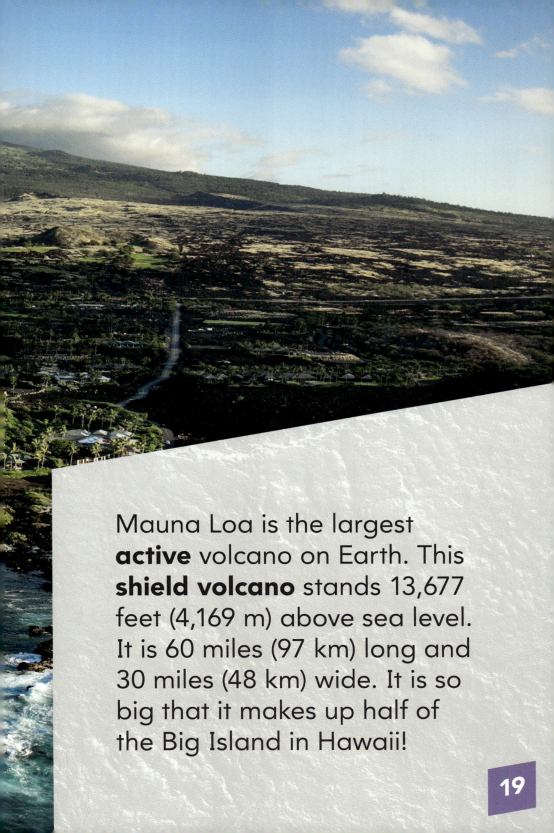

Mauna Loa is the largest **active** volcano on Earth. This **shield volcano** stands 13,677 feet (4,169 m) above sea level. It is 60 miles (97 km) long and 30 miles (48 km) wide. It is so big that it makes up half of the Big Island in Hawaii!

Mount Saint Helens is an **active** volcano in Washington state. Its 1980 eruption was so big that it was heard in many of the surrounding states! It is the deadliest volcanic event in US history, taking 57 lives.

20

# More Famous Volcanoes

- **Krakatoa, Indonesia** – its 1883 eruption was one of the most violent in history, killing 40,000 people

- **Mount Fuji, Japan** – the tallest mountain in Japan and popular tourist spot that last erupted in 1708

- **Mount Pelée, Martinique** – erupted in 1902 wiping out the town of Saint Pierre and 30,000 of its residents

- **Mount Toba, Indonesia** – its eruption about 75,000 years ago was one of the largest ever, causing a **volcanic winter** that lasted around 10 years

# Glossary

**active** – a volcano with a recent history of eruptions.

**ash cloud** – a cloud of something, such as small pieces of rock, that appears during a violent and explosive volcanic eruption.

**debris** – scattered pieces left after something has been destroyed.

**famous** – well known.

**recorded history** – written or documented history.

**shield volcano** – a broad domed volcano with gently sloping sides.

**volcanic winter** – the cooling of Earth's surface after a large volcanic event due to ash and other materials in the air that block out the sun.

# Index

dangers 7, 10, 12, 20
Earth 4, 19
effects 7, 10, 12, 14, 20
Hawaii, USA 16, 19
Indonesia 12
Kīlauea 16
Mauna Loa 19
Mount Etna 9
Mount Pinatubo 10

Mount Saint Helens 20
Mount Tambora 12
Mount Vesuvius 14
Philippines 10
Pompeii, Italy 14
Sicily, Italy 9
Washington, USA 20

# Online Resources

**Booklinks**
**NONFICTION NETWORK**
FREE! ONLINE NONFICTION RESOURCES

To learn more about famous volcanoes, please visit **abdobooklinks.com** or scan this QR code. These links are routinely monitored and updated to provide the most current information available.